Genre Nonfiction

W9-AWH-970

? Essential Question

What is energy and how can it change?

When Energy Changes

by ADRIENNE SCHURE

Chapter 1

Energy Basics

What Is Energy?

What is **energy**? In science, energy is "the ability to do work." Cleaning your room or doing homework is work for you. Scientists have a special meaning for work, though. Work is using **force** to move something. A force can cause a physical change.

Riding a bicycle uses energy. That means it is work!

Energy comes in many different forms. Light and sound are both forms of energy. Heat, motion, and electricity are forms of energy, too. Chemical reactions give energy. Magnets have another form of energy. A form of energy can change into almost any other form of energy!

Energy comes in many forms. It can also change from one form to another.

(tl) ©DLPhoto/Alamy; (tc) ©PhotoAlto; (tr) ©Image Source, all rights reserved.; (bc) Tim Keatley/Alamy; (bl) Rodrigo Torres/Glow Images; (br) Jose Luis Pelaez Inc/Getty Images

Where Does Energy Come From?

A cat jumps on a table and knocks over a glass. The glass falls to the floor and breaks. It makes a loud noise. There are many forms of energy at work. The cat uses stored chemical energy to jump onto the table. The cat puts force on the glass. Energy goes to the glass, and the glass falls over. It rolls to the edge of the table. It falls to the floor. This is motion energy. The glass breaks and makes sound energy.

It all starts with the cat. Where does the cat get the energy to jump on the table?

Almost all energy on Earth comes from the Sun. Sunlight is the energy that comes from the Sun. Plants store this energy and change it into chemical energy. Plants use chemical energy to live and grow. People and animals eat plants and other animals. Then the chemical energy from the plants moves to the people and animals. So the cat's energy starts with the Sun!

The Sun is Earth's energy source.

Where Does Energy Go?

Physics is the science of matter and energy. A law of physics says that energy cannot be created or destroyed. But energy *can* change from one form to another. Energy can also move. Gasoline has chemical energy. A car changes this chemical energy into mechanical energy. This causes the engine to run and the car to move.

Nuclear energy

Chemical energy

Some forms of energy

Electrical energy

Light energy

Sound energy

Kinetic energy

Other forms of energy

The car does not use all the energy from the fuel to move. A lot of energy changes to heat. This thermal energy is wasted energy. The car gets hot, but it does not use this energy.

This book is about changes in energy. We will learn how one form of energy changes to another form.

Chapter 2
Changing Energy from the Sun to Make Cars Go

Every few years there is a contest for college students. They try to design and build special cars. These cars use energy from the Sun. Sun power is called **solar energy**.

The cars are part of a race. In 2014, the race started in Austin, Texas. It ended in Minneapolis, Minnesota. The cars used solar energy to travel 1,824 kilometers (1,134 miles)!

Teams of students want to build the best car. A car needs to collect energy from the Sun to win. Then the car needs to change solar energy to another form of energy. This energy makes the car move.

One of the solar cars in a race

What Is Electric Current?

Matter is made of tiny parts called atoms. **Electrons** are parts of atoms. Scientists can make electrons move from one atom to another. This movement, or flow, of electrons is called an **electric current**.

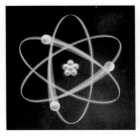

Electrons fly around the center of the atom.

Making the Change

These cars all have special solar cells that change the Sun's energy into electricity. They use the Sun's energy to make an electric current.

The electric current gives power to the car's motor. The motor moves the wheel and other parts and makes the car go.

What happens when the car is not moving? The electric current goes to the **batteries**. The current gets saved as chemical energy. The batteries hold the chemical energy until it is changed into electrical energy.

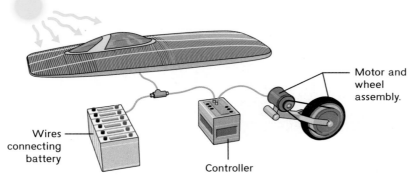

Wires connecting battery

Controller

Motor and wheel assembly.

Flow of Energy in a Solar Car

This car uses solar energy.

The Good and Bad of Solar Energy

Using solar energy instead of gasoline for cars can have advantages. These cars do not cause pollution. Also, there will always be solar energy. Gasoline is made from oil. There will not always be oil because there is a limited supply of oil on Earth.

Solar energy does have some problems. Sometimes there is no sunlight. A car's solar cells can get a lot of energy on a sunny day. The car can store some of this energy in its battery. But that is not the case on a rainy day or at night.

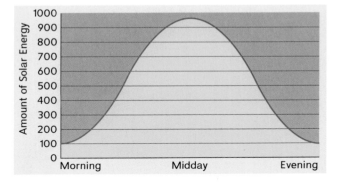

The amount of energy from sunlight is highest in the middle of the day.

The Sun's energy is weaker in winter. Cars can use energy in their batteries. The batteries run out of energy if the Sun does not shine for a few days. Then the car stops! It has no source of energy.

We have not found a good way to store solar energy. People are trying to solve this problem. Then solar energy could power cars and heat homes.

These solar panels are collecting sunlight. The energy will be changed to electricity to use in homes and businesses.

Darren Baker/Alamy

Chapter 3

Light Up the World

Who Turned on the Lights?

Michael Faraday was a scientist in England. Faraday did many experiments. He moved a copper wire through a **magnetic field**. This made electric current flow through the wire. A magnetic field is the area around a magnet with a pushing or pulling force.

This picture shows Michael Faraday (1791–1867) doing one of his experiments with magnetism and electricity.

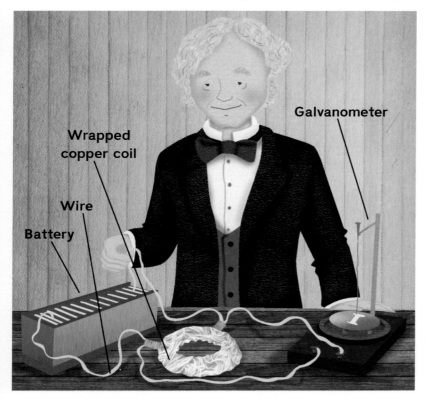

Galvanometer

Wrapped copper coil

Wire

Battery

Faraday's experiments helped us with electricity. We use electricity in many ways today because of his work. A lot of our electricity comes from power plants. A power plant is like a factory that makes electricity. Some power plants use moving water to make electricity. The force of water moves **turbines**. A turbine is a machine with large blades like a fan. Water moves over the blades and turns them. This makes electric current.

What happens to the electric current? It flows through wires. The path that the current takes is called a **circuit**. The current reaches your home and then you can control it. You turn on a switch and current flows through a circuit in your home. You turn off a switch and the circuit is stopped.

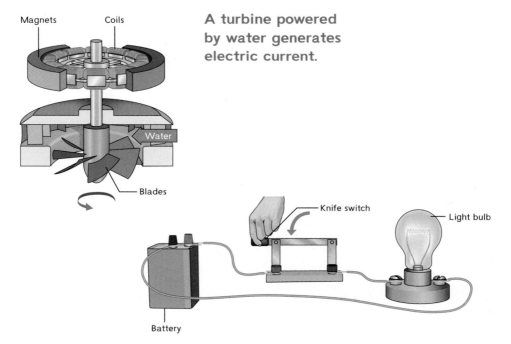

Magnets   Coils

A turbine powered by water generates electric current.

Water

Blades

Knife switch   Light bulb

Battery

Electric current flows from the battery through a circuit. When the bar is raised, the circuit is broken and the light goes off.

Turning Up the Heat

Think about the light bulbs in your home. Incandescent light bulbs contain a thin wire inside called a filament. The filament completes an electric circuit and makes an incandescent light bulb shine.

Here is how these light bulbs work. There are tightly packed atoms in the thin wire. Electrons push their way through the atoms of the wire. This produces heat. The wire becomes hotter and hotter. It first glows red, but then it glows white. The white-hot wire makes the light bulb bright. It also makes the light bulb hot.

Flowing electrons produce heat. Heat causes the light bulb's filament to glow white hot.

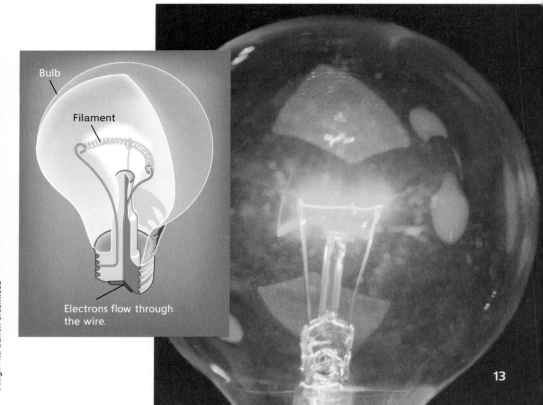

Bulb

Filament

Electrons flow through the wire.

Turning Down the Heat

Light bulbs use heat to make light. It takes a lot of electrical energy to make heat. But we want light bulbs to make light, not heat. We do not use heat in the light bulb. The energy used to make heat is wasted. Only a little electrical energy changes to light.

A fluorescent light bulb makes light in another way. There is an electric current that flows through a circuit. A fluorescent bulb does not have a filament, though. It has gases inside the bulb instead. Electric current changes the gases. The gases send out light. This is called UV (ultraviolet) light. The UV light hits powders on the inside of the glass tube. Then the powders give off white light. A fluorescent bulb changes electricity to light energy.

Fluorescent light bulbs use less energy than those with filaments. They do not give off as much heat. A light bulb with a filament gets very hot when it is on. A fluorescent bulb stays much cooler.

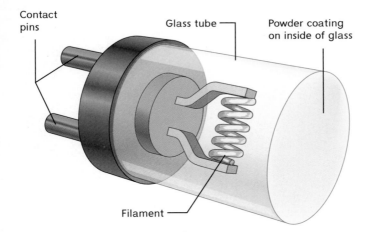

Contact pins

Glass tube

Powder coating on inside of glass

Filament

Parts of a Fluorescent Light Bulb

Chapter 4
Energy You Can Carry

You probably use batteries every day. They are in cell phones, portable music players, toys, and flashlights. Batteries store energy in the form of chemical energy.

You use batteries in many ways.

How Batteries Work

Inside batteries are metals and chemicals. These metals and chemicals make electricity flow when the circuit is complete. You turn on the camera or toy to complete to the circuit. Electric current can flow. A chemical reaction changes stored energy into heat, light, sound, or another form of energy.

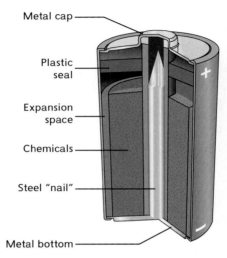

Metal cap

Plastic seal

Expansion space

Chemicals

Steel "nail"

Metal bottom

Different Types of Batteries

All batteries have metals and chemicals. They make a force that moves electrons. Different types of batteries use different chemicals and metals. Some batteries make more power than others.

A Battery for Every Need	
Type of Battery	Where It Is Used
Alkaline/AAA, AA, C, D	Toys, simple games, small flashlights
Lead-acid	Automobiles
Lithium	Camera flash
Lithium-ion	Laptop computers, cell phones, digital cameras

Chapter 5
Sound Gets Around

Your favorite band is in a recording studio recording a new song. Each member of the band has a microphone. Each microphone connects to something called a soundboard. This controls the sound. The band plays. The instruments send sound waves to the microphones. The microphones change the sound into electrical energy. This energy gets changed to a digital code. It gets stored in a computer or on a disk. Then you can store it on your music player or computer.

Inside a Microphone

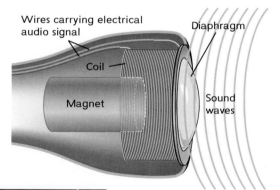

Wires carrying electrical audio signal

Diaphragm

Coil

Magnet

Sound waves

A recording engineer uses a soundboard in a recording studio.

How can we hear the music? The energy changes again. It becomes sound waves. Music players and computers have speakers. Speakers change the energy. When you use them to play music, the digital code changes into an electric signal. Speakers have a thin, flexible sheet inside. Electrical energy makes the thin sheet move back and forth. These **vibrations** are sound waves you hear.

Super Sound

Dolphins swim in the ocean. Dolphins use sound to find out about the ocean. A dolphin makes clicking noises as it swims. These noises are sound waves. They go through the water. A sound wave hits an object and bounces back to the dolphin. These are called echoes. A dolphin can understand the object from the echoes. Echoes tell a dolphin if something is hard or soft. A dolphin knows if it is near or far, what shape it is, and how fast it is moves.

Martin Ruegner/Getty Images

Using Energy

Take a walk. Play music. Turn on the lights. Talk on your cell phone. Ride in a car. You change energy when you do these things.

It is important to use energy wisely each day. Do not waste energy! In the future, we will use energy in different ways. We will find new ways to change energy from one form to another. Yet each way will begin with the Sun.

This kind of stove allows people to cook food by changing light energy to heat.

Respond to Reading

Summarize

Use important details from *When Energy Changes* to sum up what you read. Your graphic organizer may help you.

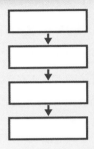

Text Evidence

1. How can you tell that *When Energy Changes* is a nonfiction text? GENRE

2. How might things we learn in the future help us waste less energy? Why might this be important? INFER

3. Find the word *flexible* on page 18. What does this word mean? Which context clues helped you figure it out? CONTEXT CLUES

4. We may be able to store more energy in the future. How might this help people? Write a story about what life might be like in 100 years. WRITE ABOUT READING

Compare Texts

Read about how energy can change form. Learn how energy from the Sun can help in a race.

As Fast As Light?

"Can we do it?" She made her seat belt tight. "Can we take the lead in the race? Can we take first place?" Malia knew that as her team's driver she would do her best.

Malia put her phone in the holder just above the steering wheel. A smile came to her face as she thought about how Jacob had made an app that let her control every part of the car.

The solar collectors on the car took the energy that came all of the way from the Sun. They then created the electricity that made the car run. Those same collectors charged the battery in her cell phone and the car battery.

Malia scratched her left ankle. She knew why she drove the car. She was small and light. No one of the others could have scratched their ankle. There wasn't enough room for any of them to fit in the light and hard seat. The car was a solar car. It had to be small and light.

Malia asked for the data screen. "The sun is up at 6:15. Power levels, please?" A screen told her that the car could run on the batteries for 35 minutes before the solar panels kicked in. A careful calculation told Malia that she could start at 6:00. "Time please!" Malia saw the time tick off and thought, "May the best girl win!"

Make Connections

In *As Fast As Light?* What do Malia and her team do to make a car powered by the Sun?

Glossary

battery *(BAT-uh-ree)* a container filled with chemicals that makes electrical power *(page 8)*

circuit *(SUR-kit)* a closed path for electric current to travel through *(page 12)*

electric current *(i-LEK-trik KUHR-uhnt)* the flow of electrons through a wire path *(page 8)*

electron *(i-LEK-tron)* a particle that moves around the center of an atom *(page 8)*

energy *(EN-uhr-jee)* the ability to do work *(page 2)*

force *(FORS)* a push or pull on something that causes a physical change *(page 2)*

magnetic field *(mag-NET-ik FEELD)* the space around a magnet where the magnetic force pushes and pulls *(page 11)*

physics *(FIZ-iks)* the science of energy and matter *(page 5)*

solar energy *(SOH-luhr EN-uhr-jee)* energy from the Sun that can be used directly for making heat or electricity *(page 7)*

turbine *(TUHR-bine)* a machine powered by water passing through blades of a wheel, turning them *(page 12)*

vibration *(vye-BRAY- shuhn)* quick movement back and forth *(page 18)*

Index